How to use this book

A sample page

INSTRUCTION
What your child needs to do for the activity.

TITLE
The page title describes the skill your child will learn in these pages.

FUN ILLUSTRATIONS
Specially drawn illustrations which are fun, interesting and drawn at the right pedagogical level for your child.

COLOURFUL BORDERS
The page borders make each page as attractive as possible to stimulate your child.

EXAMPLE
The first one is done for you so you can show your child exactly what to do.

LOTS OF PRACTICE
Two pages where your child can practise and repeat the same skill to master it.

STICKERS
Place a sticker on each page as your child finishes.

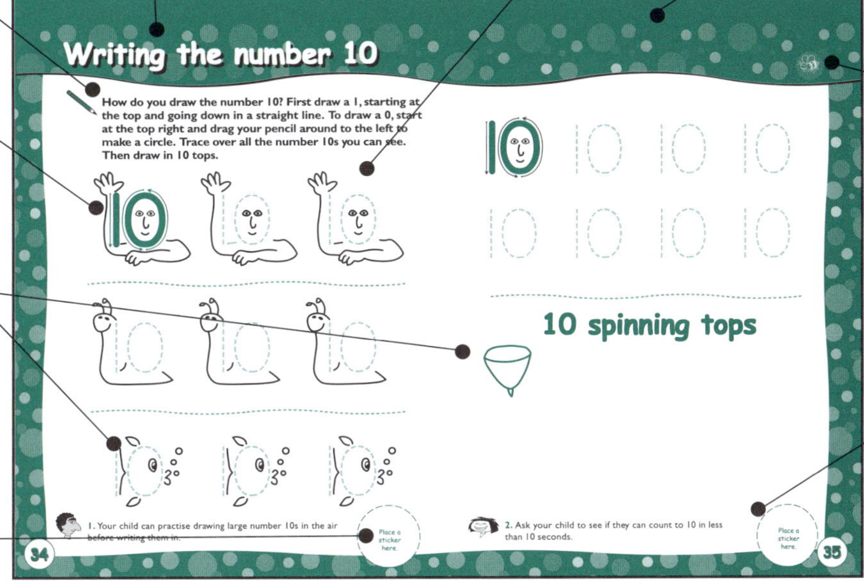

WHO'S HIDING?
In each book, a little creature appears in the border of every double page so your child can have fun trying to find it.

EXTRA ACTIVITIES
Extra activities you might want to do with your child to further reinforce the skill or simply make it more enjoyable.

Step-by-step learning

STEP ONE — **Read** out the title of the activity page to your child.

STEP TWO — **Explain** the skill and show your child the example already done. **Make sure** they understand what to do. Your child will then have at least two pages to practise that same skill.

STEP THREE — **Help** your child put a **sticker** on the bottom of each page as they complete it.

Remember to be patient, encouraging and positive with your child, even when minor mistakes are made!

How to hold a pencil

It is important that you help your child hold his or her crayon or pencil in the correct way as shown here to ensure your child develops the right technique early on.

Reviewing numbers 1–5

Look at the cards hanging on the lines. How many pictures are on the first card? Write the matching number on the middle card, and draw the same number of triangles in the last card.

2

5

1. Ask your child to name the shapes they are drawing in the bottom row.

Place a sticker here.

2. Point to a number. Ask your child to collect that many objects from the room.

Place a sticker here.

Making patterns of 6

Pick two different coloured pencils. Colour the first 6 shapes in one colour, then the next 6 shapes in the other colour.

1. Ask your child to count the fish on this page.

Place a sticker here.

2. You can help your child make repeating patterns of 6 with objects around the house or with movements such as 6 jumps, 6 steps, 6 jumps and so on.

Place a sticker here.

Drawing groups of 6

Draw matching pictures or shapes so there are 6 things in each group.

6 balloons

6 legs

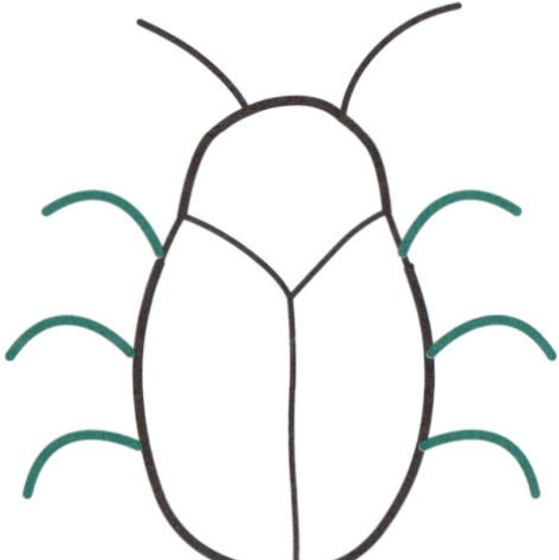

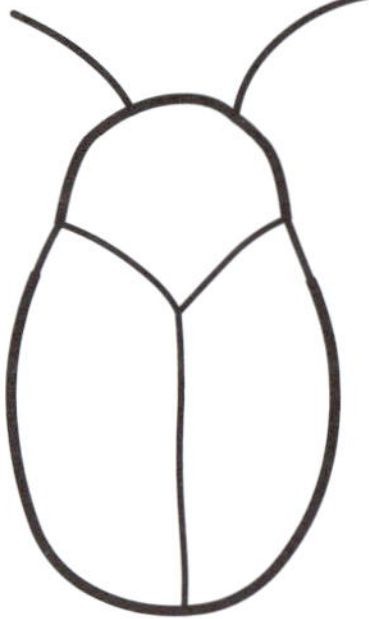

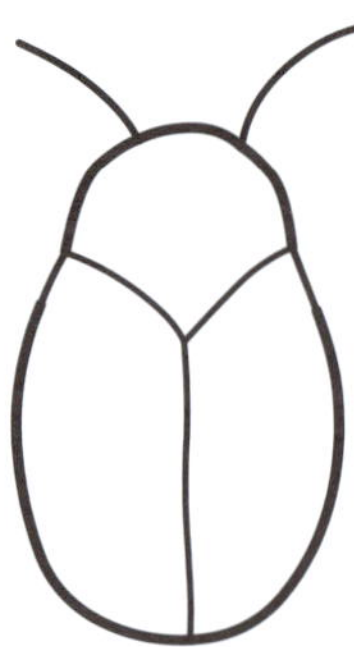

1. See if your child can see any other groups of 6 around you in the home. Are there 6 cups that look the same? Or 6 forks or spoons?

Place a sticker here.

6 spots

6 triangles

2. Ask your child to hold up 6 fingers and count them.

Place a sticker here.

Writing the number 6

How do you draw the number 6? Start at the top and drag your pencil down to the left and around. Trace over all the number 6s you can see. Then draw in six snakes.

1. Your child can practise drawing large number sixes in the air before writing them in.

Place a sticker here.

6 snakes

2. See how many number 6s your child can find around you when you next go out. Are there any house numbers with the number 6 in them? Shop signs? Road signs?

Place a sticker here.

Making patterns of 7

Pick two different coloured pencils. Colour the first 7 shapes in one colour, then the next 7 shapes in the other colour. Then go back to the first colour if there are any pictures left.

1. Ask your child to hold up 7 fingers and count them.

Place a sticker here.

2. Ask your child to point to the group which has the *most* pictures in it. Check by counting.

Place a sticker here.

Drawing groups of 7

Draw matching pictures so there are 7 things in each group.

7 kittens

7 fish

1. Encourage your child to count whenever you can. How many steps can they take across a room? How many spoons are laid out on the table? How many birds can they see outside?

Place a sticker here.

7 glasses

7 gum leaves

2. Are there 7 kittens, 7 fish, 7 glasses of water and 7 gum leaves? Check by counting.

Place a sticker here.

Writing the number 7

How do you draw the number 7? Start at the top left. Drag your pencil across to the right and down to the left. Trace over all the number 7s you can see. Then draw in 7 sheep.

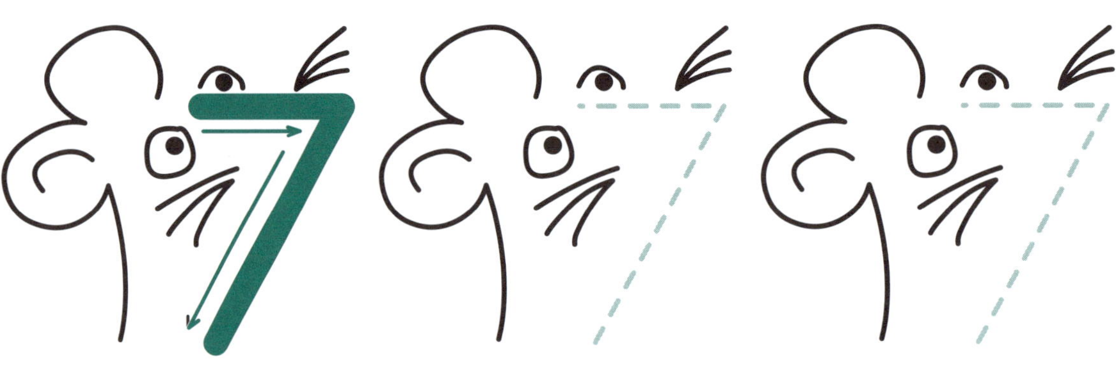

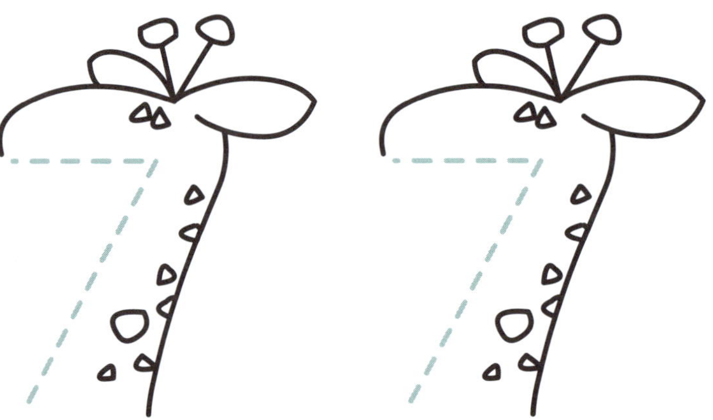

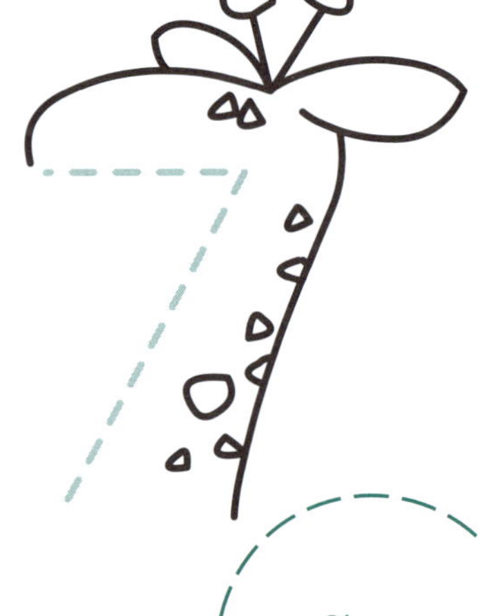

1. Your child can practise drawing large number sevens in the air, before writing them in.

7 sheep

2. Ask your child to stamp their feet 7 times and clap their hands 7 times. What else can they do?

Place a sticker here.

Reviewing numbers 1–7

Look at the cards hanging on the lines. How many pictures are on the first card? Write the matching number on the middle card, and draw the same number of circles in the last card.

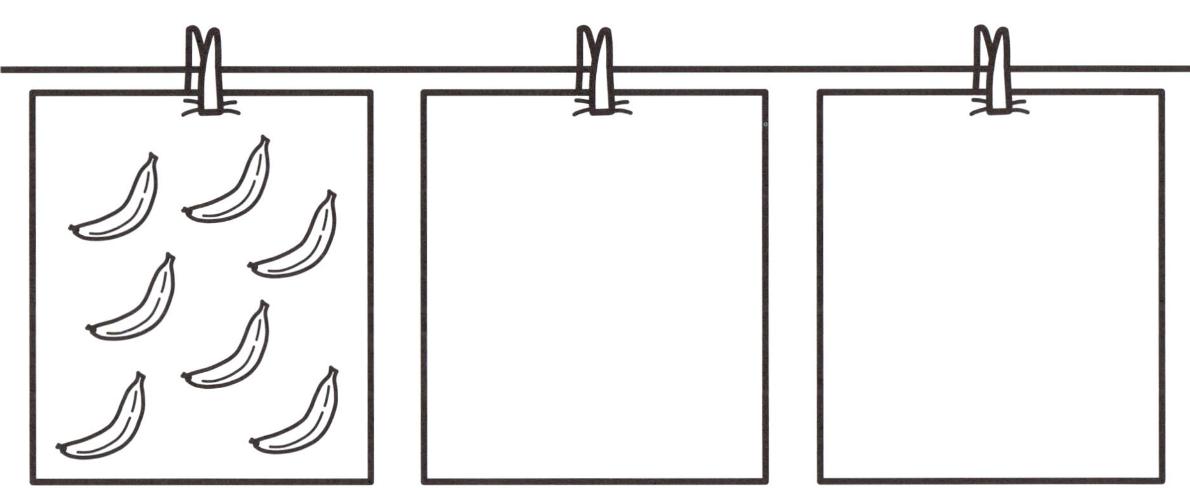

1. Find 6 forks and match them to 6 spoons.

Place a sticker here.

2. Find 7 cups and match them to 7 saucers.

Place a sticker here.

Making patterns of 8

Pick two different coloured pencils. Colour the first 8 shapes in one colour, then the next 8 shapes in the other colour.

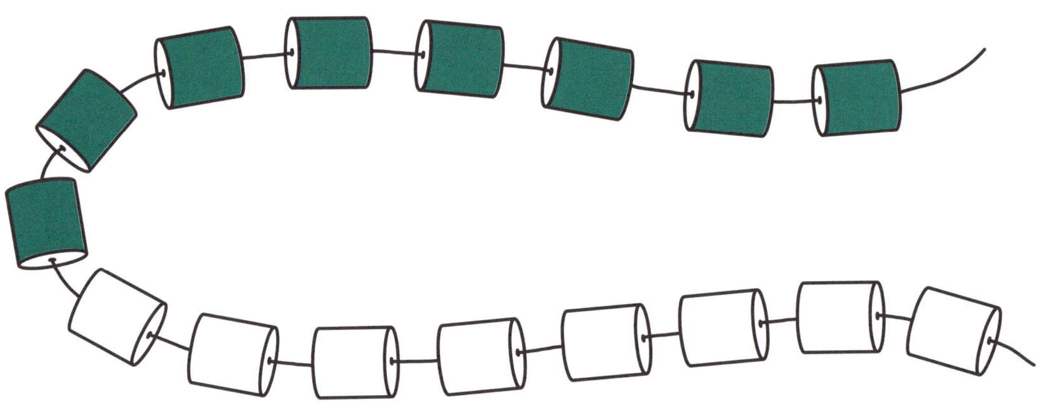

1. Ask your child to hold up 8 fingers and count them.

Place a sticker here.

2. Ask your child to nod their head 8 times, and shake their head 8 times. What else can they do?

Place a sticker here.

Drawing groups of 8

Draw matching pictures so there are 8 things in each group.

8 hairs

8 bananas

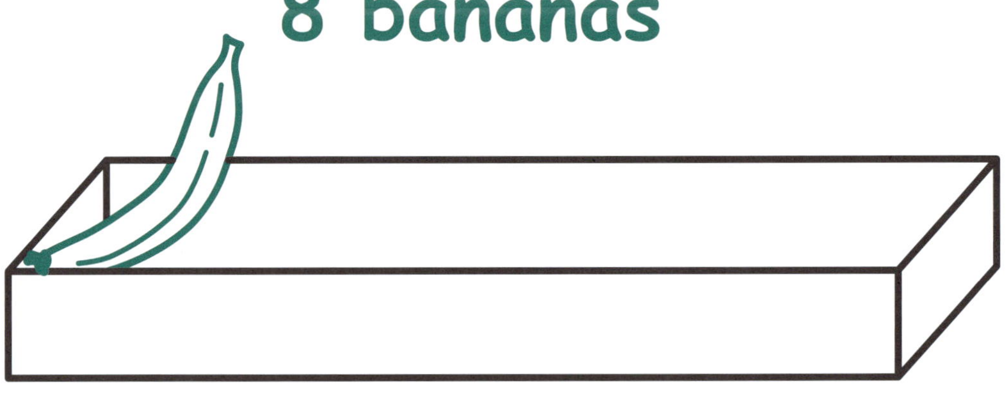

8 wheels

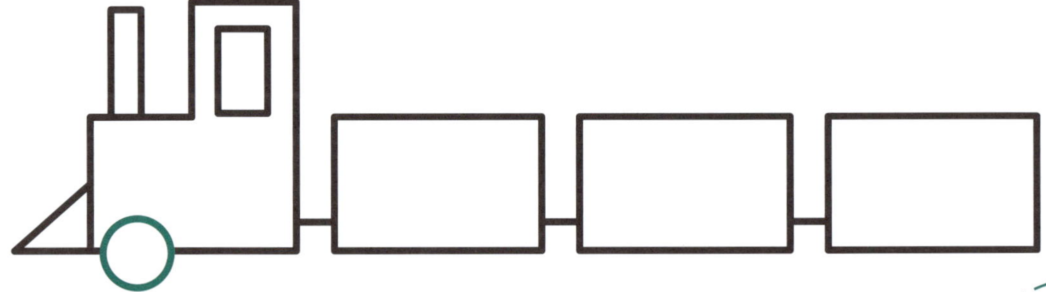

1. Help your child count to 8 by counting along with a simple song or piece of music.

Place a sticker here.

8 carrots

8 spots

8 squares

2. Ask your child to show you their 8th finger and their 8th toe.

Place a sticker here.

Writing the number 8

How do you draw the number 8? Start at the top right. Curve your pencil around to the left and down to the bottom right in a smooth curve, then back up to join where you started. Trace over all the number 8s you can see. Then draw in 8 suns.

1. Your child can practise drawing large number 8s in the air before writing them in.

Place a sticker here.

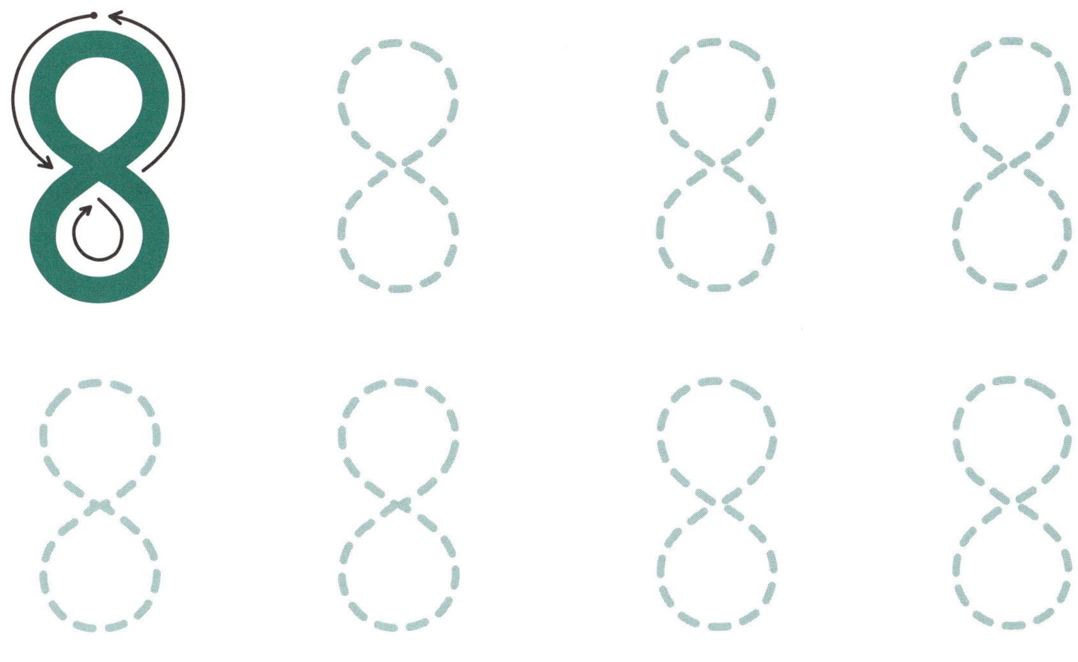

8 suns

2. Draw 8 large squares on paper. Ask your child to draw 8 sticks inside each square.

Place a sticker here.

Making patterns of 9

Pick two different coloured pencils. Colour the first 9 shapes in one colour, then the next 9 shapes in the other colour. Then go back to the first colour if there are any pictures left.

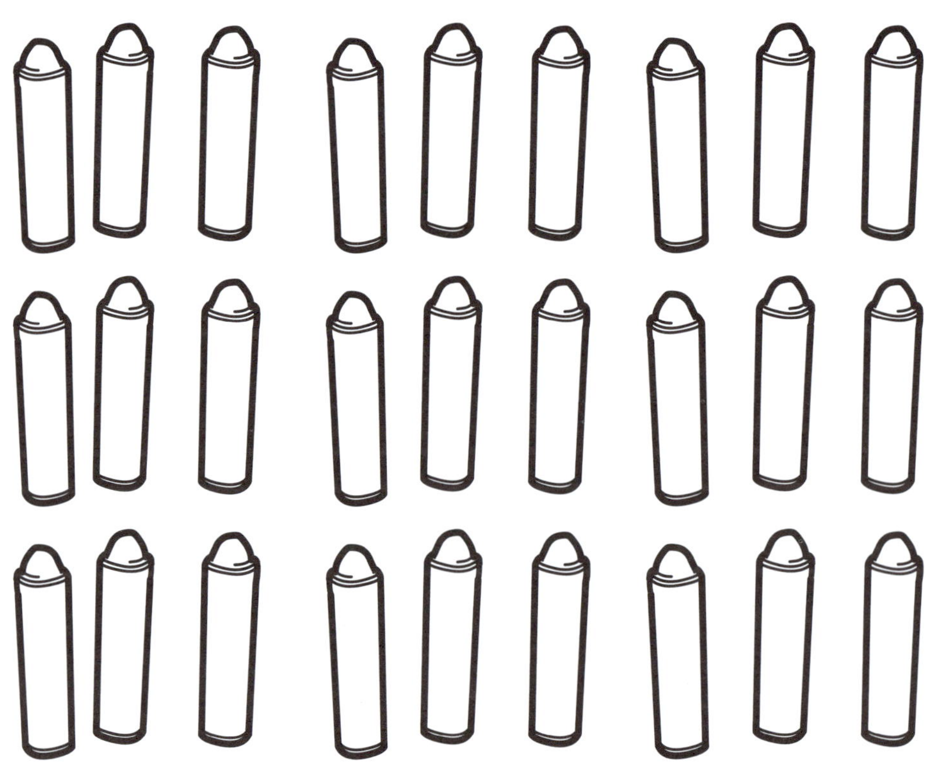

1. Ask your child to hold up 9 fingers and count them.

Place a sticker here.

2. Can your child count how many sailboats are on this page?

Place a sticker here.

Drawing groups of 9

Draw matching pictures so there are 9 things in each group.

9 olives

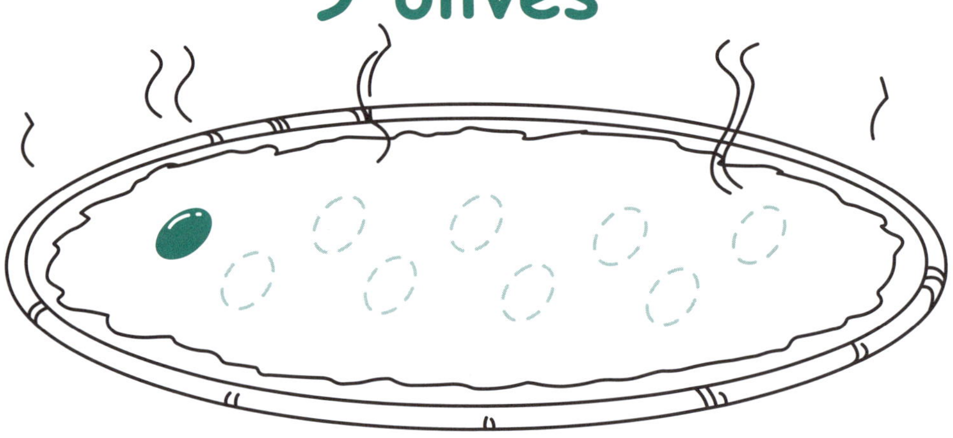

9 stripes

1. Ask your child to guess how far away they would be if they walked 9 steps. Then they can check.

Place a sticker here.

9 eggs

9 books

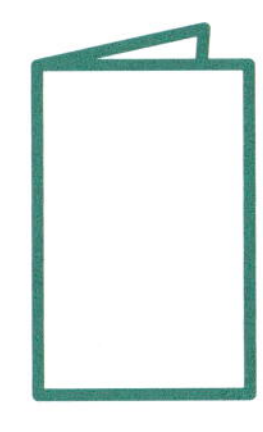

2. Ask your child to guess how long 9 books placed end-to-end would be. Then they can check.

Place a sticker here.

Writing the number 9

How do you draw the number 9? Start at the top right and drag your pencil around to the left to make a circle. Then drag your pencil down to make a stick. Trace over all the number 9s you can see. Then draw in 9 nuts.

1. Your child can practise drawing large number nines in the air before writing them in.

Place a sticker here.

9 nuts

2. Ask your child if they can stand on one leg for 9 seconds and have them count with you.

Place a sticker here.

Making patterns of 10

Pick two different coloured pencils. Colour the first 10 shapes in one colour, then the next 10 shapes in the other colour.

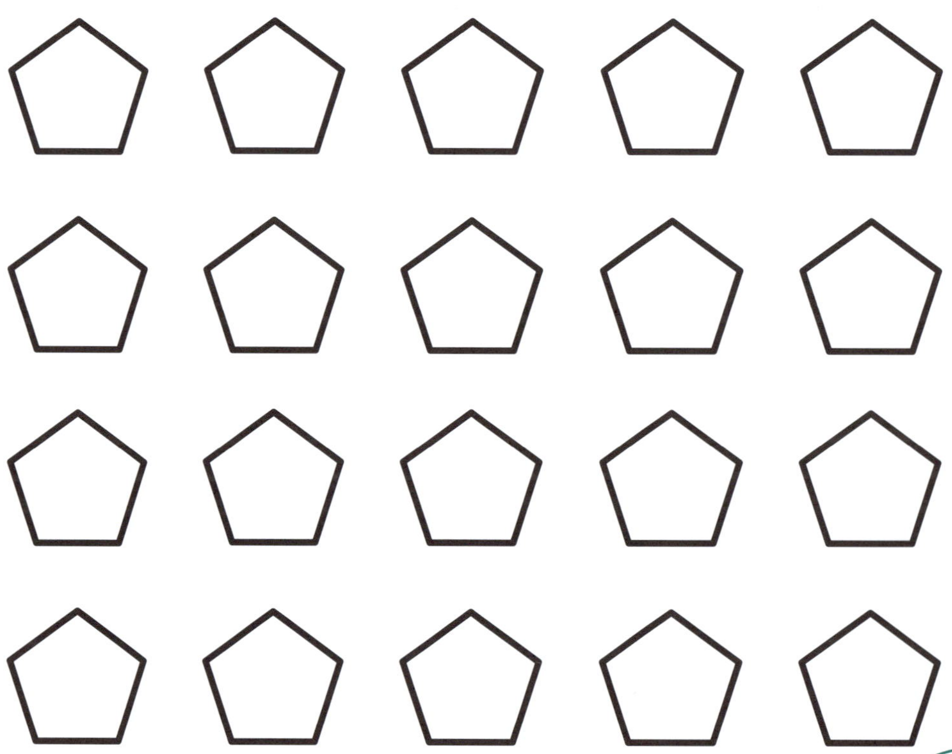

1. Ask your child to hold up 10 fingers and count them.

Place a sticker here.

2. Sing songs or rhymes which involve the number 10 to help your child count, such as 'Ten Green Bottles'.

Place a sticker here.

Drawing groups of 10

Draw matching pictures so there are 10 things in each group.

10 fingers

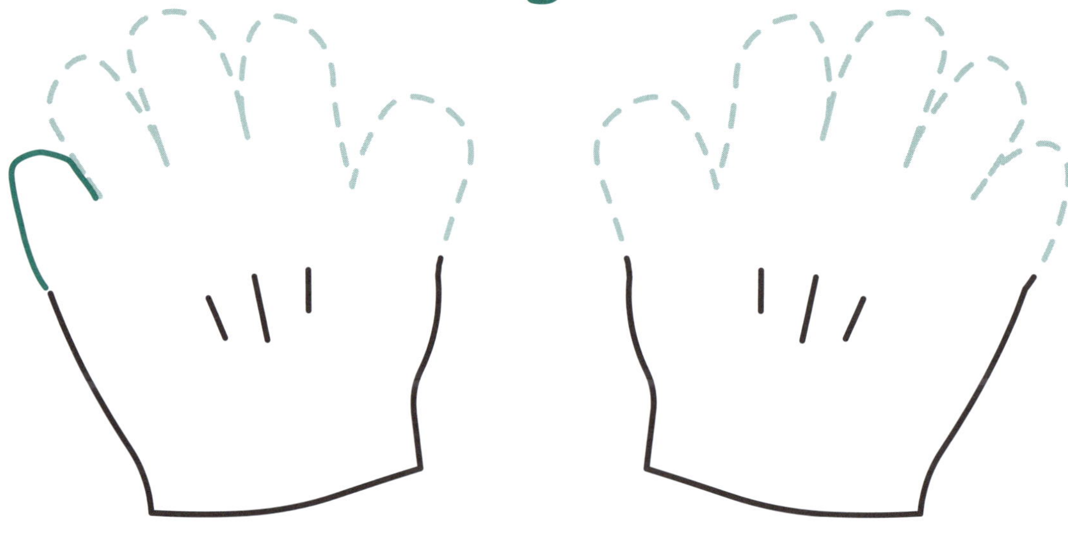

10 balls

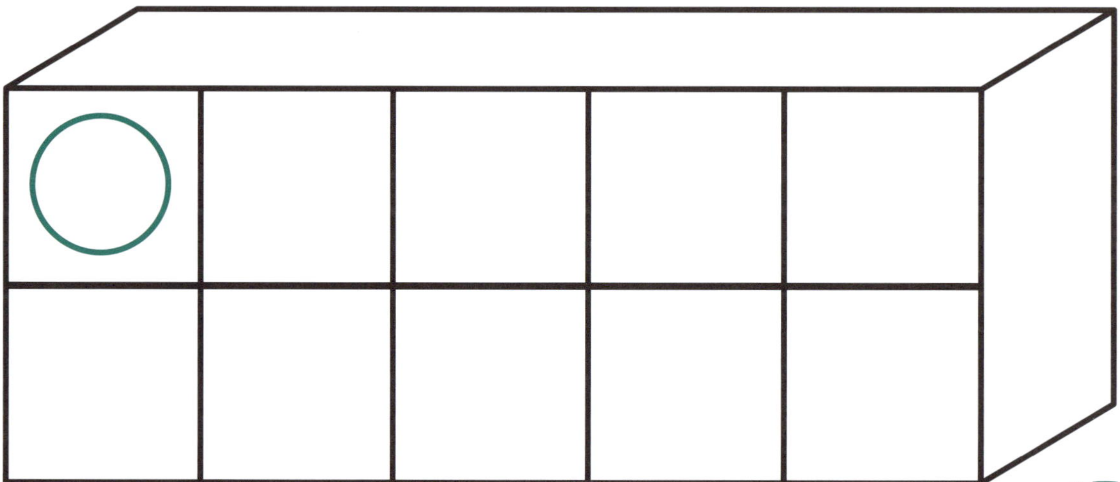

1. Ask your child to collect 10 pencils and match them to 10 pieces of paper.

Place a sticker here.

10 teeth

10 chips

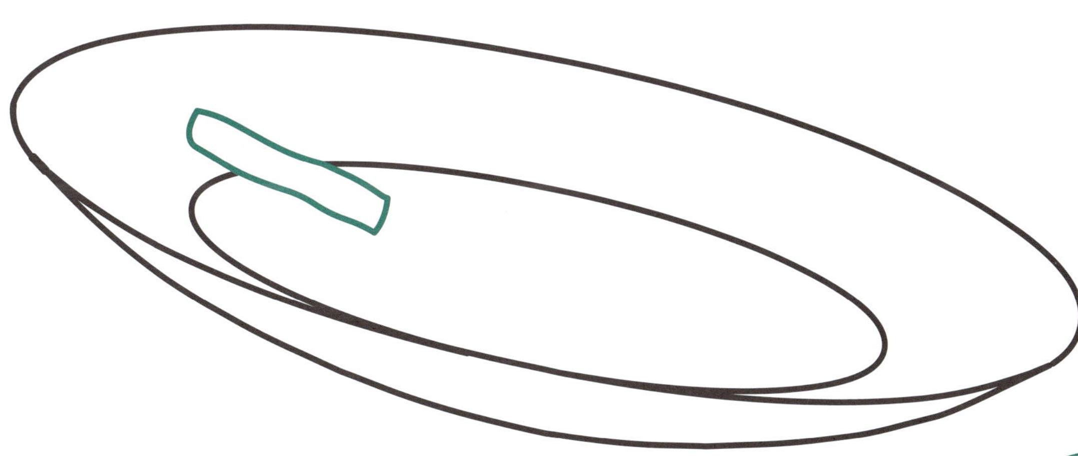

2. Ask your child if there are 10 balls in the box? 10 teeth? Count all the things on the page.

Place a sticker here.

Writing the number 10

How do you draw the number 10? First draw a 1, starting at the top and going down in a straight line. To draw a 0, start at the top right and drag your pencil around to the left to make a circle. Trace over all the number 10s you can see. Then draw in 10 tops.

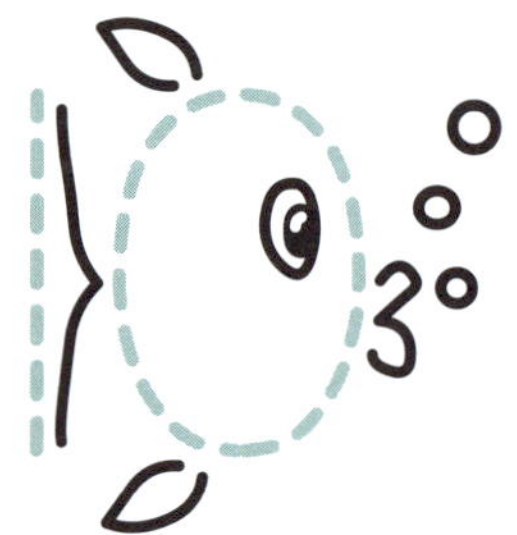

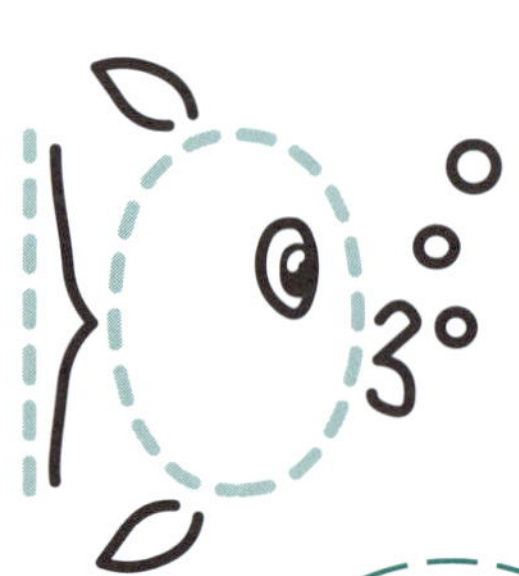

1. Your child can practise drawing large number 10s in the air before writing them in.

Place a sticker here.

10 spinning tops

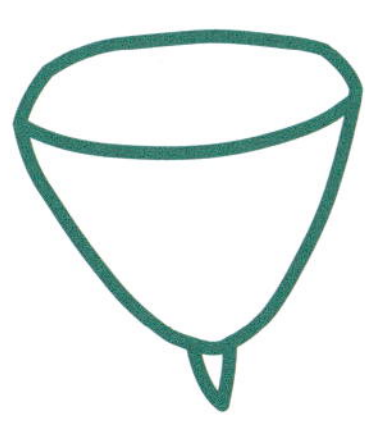

2. Ask your child to see if they can count to 10 in less than 10 seconds.

Place a sticker here.

Joining the matching sets

How many animals are in each box? Draw a line between the boxes which have the same number of animals inside them.

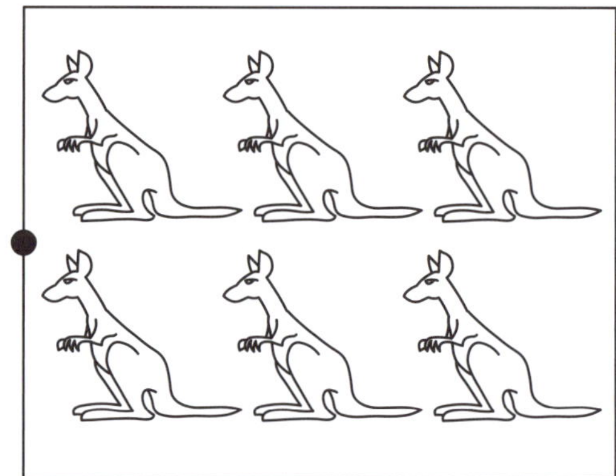

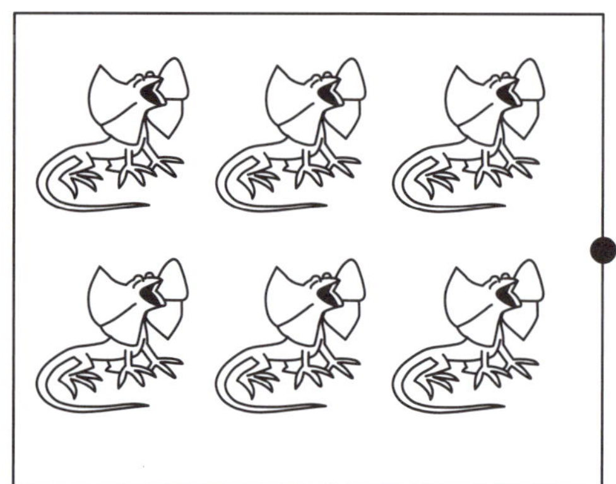

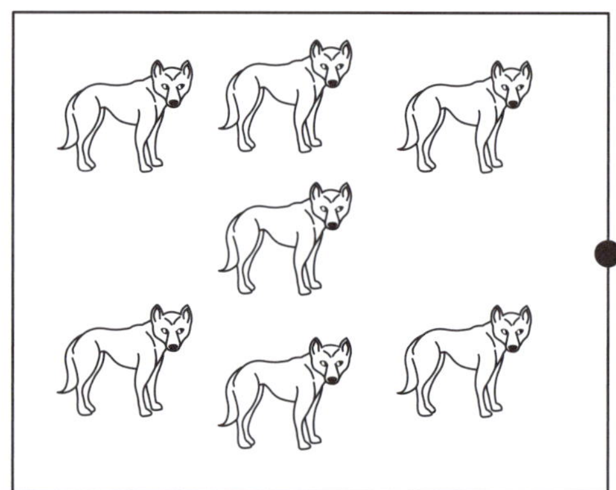

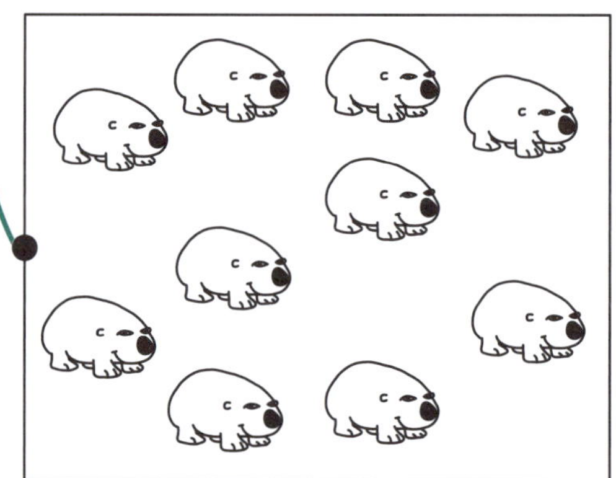

1. Ask your child if there are more echidnas than koalas. Then count them to check.

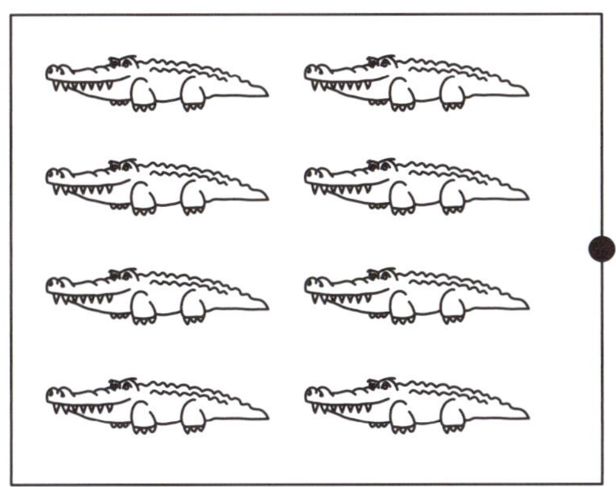

2. Ask your child if there are more tigers than snakes. Then count them to check.

Place a sticker here.

Finding the matching number

How many pictures can you see in each box? Circle the matching number.

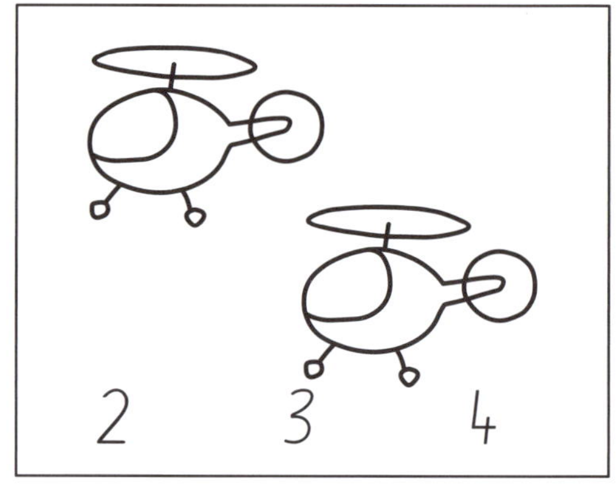

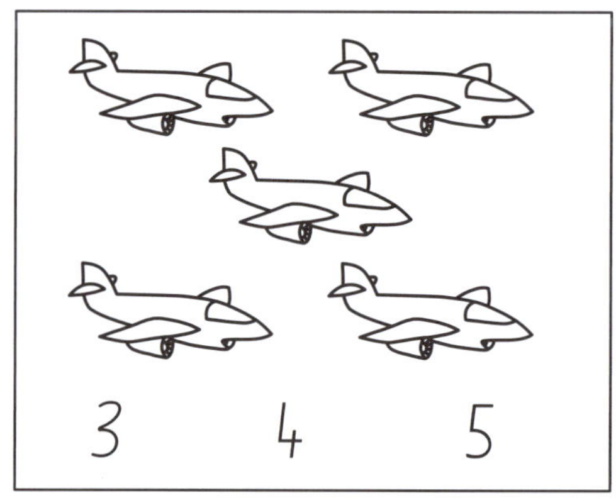

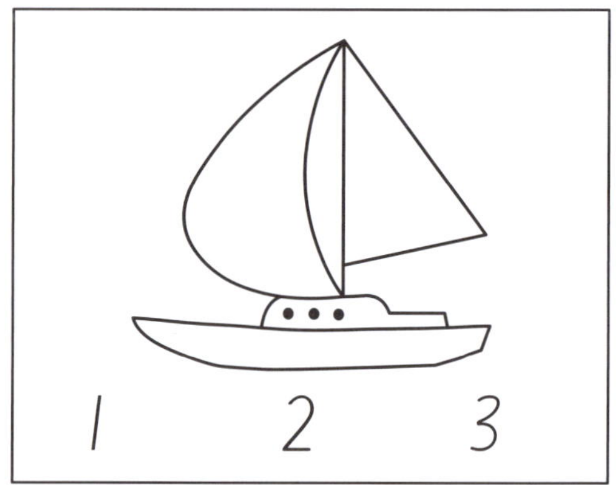

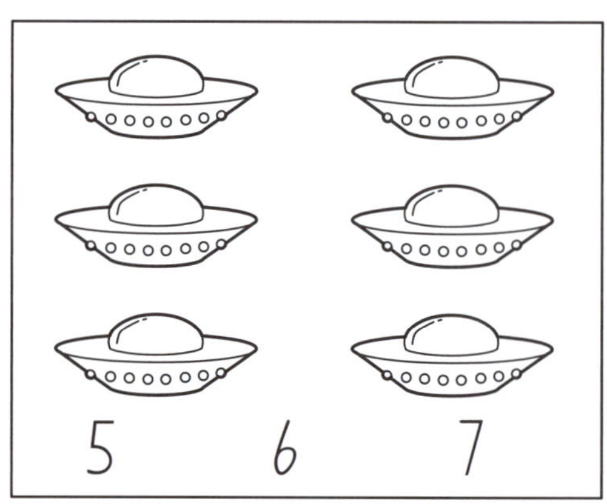

1. Ask your child if there are more jets or more spaceships. Then count them to check.

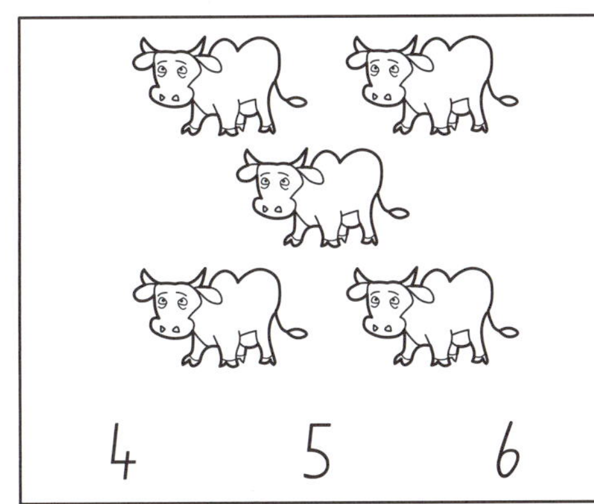

2. Ask your child if there are more ducks or more sheep. Then count them to check.

Place a sticker here.

Writing the matching number

Count the pictures in each group, then write the matching number beside them.

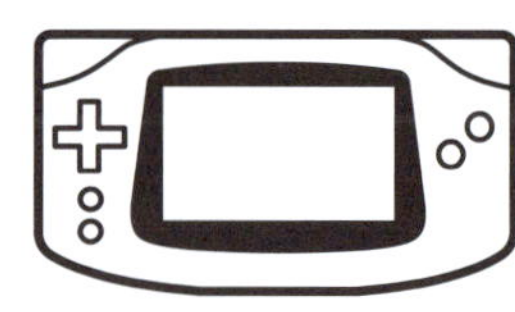

1. You can play counting games with real objects — ask your child to count a group of toys, or a pile of books, with up to 10 in the pile.

Place a sticker here.

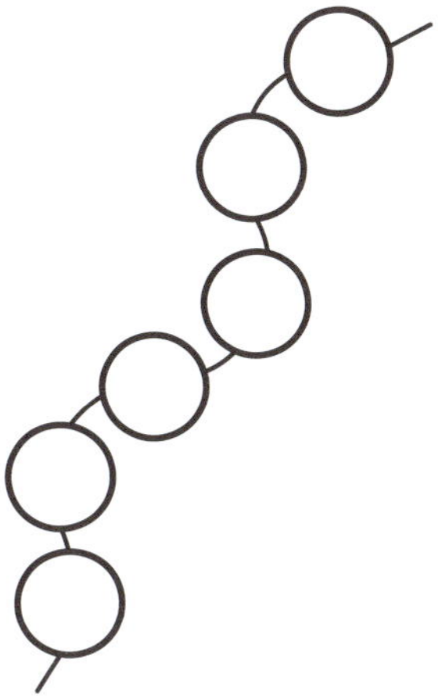

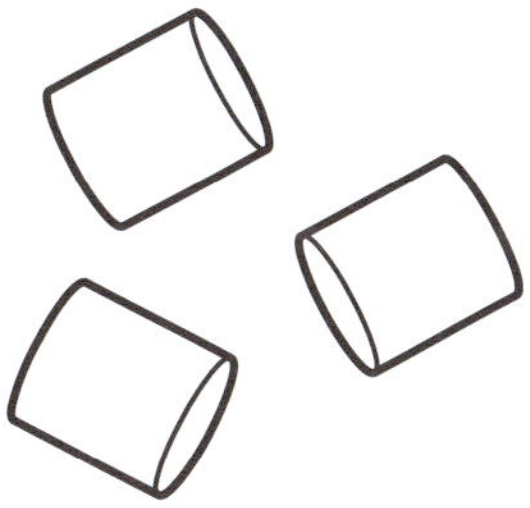

2. Ask your child to write five number 5s, seven number 7s and three number 3s.

Place a sticker here.

Drawing the matching sets

Draw fish to match the number beside each tank. Then draw spots on each cow to match the number beside each one.

5

8

6

1. Ask your child which tank has the smallest number of fish.

Place a sticker here.

2. Ask your child which cow has the smallest number of spots.

Place a sticker here.

Joining the dots

Here are two hidden pictures. Join the sets of numbers from 1 to 10 to find each hidden picture. Start at number 1 each time.

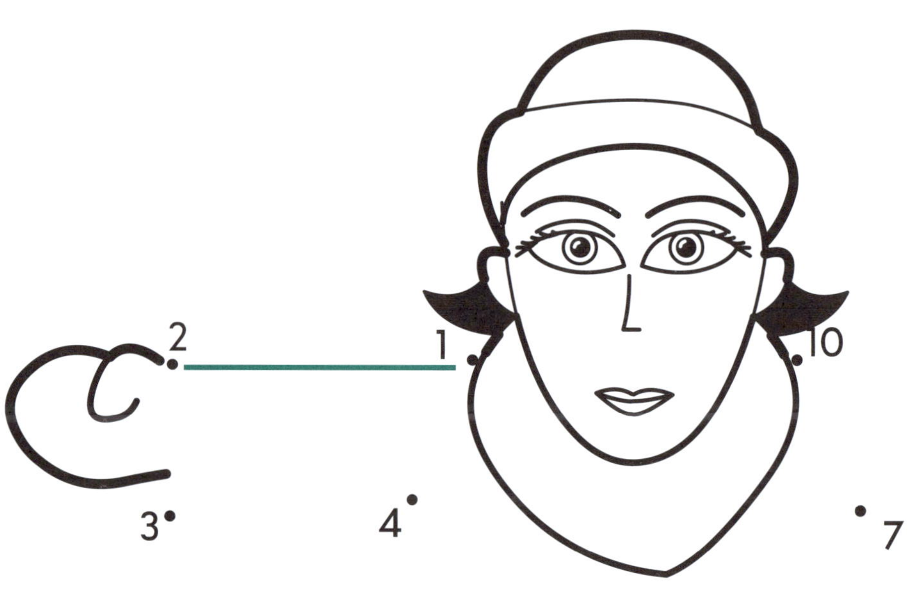

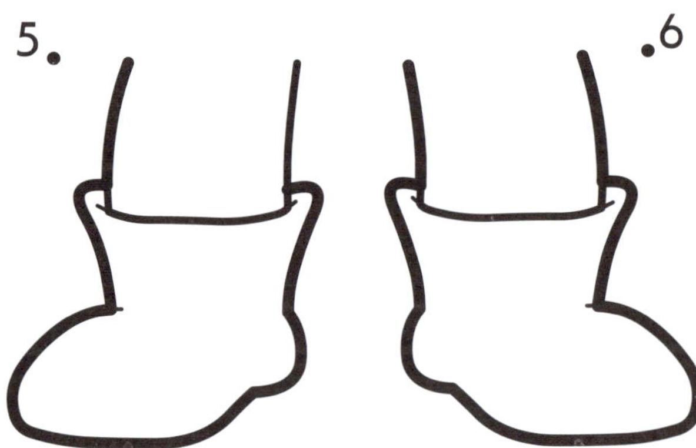

1. Try putting the numbers 1 to 10 on separate cards, shuffle them, and ask your child to put them in the correct order.

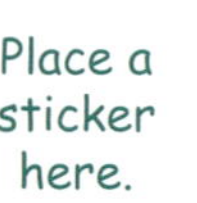

1

2

3

4

5

6

7

8

9

10

2. Where else can your child see the numbers 1 to 10 in order? On the telephone?

Place a sticker here.

Colouring by numbers

Pick 10 different coloured pencils or crayons. Colour in the circles at the top of the page, then colour the pictures to match.

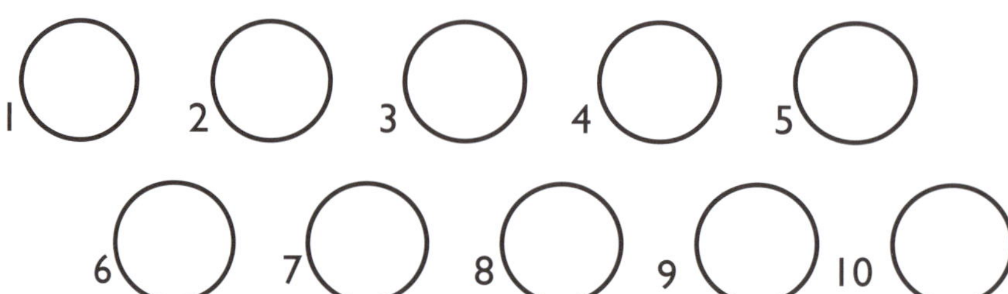

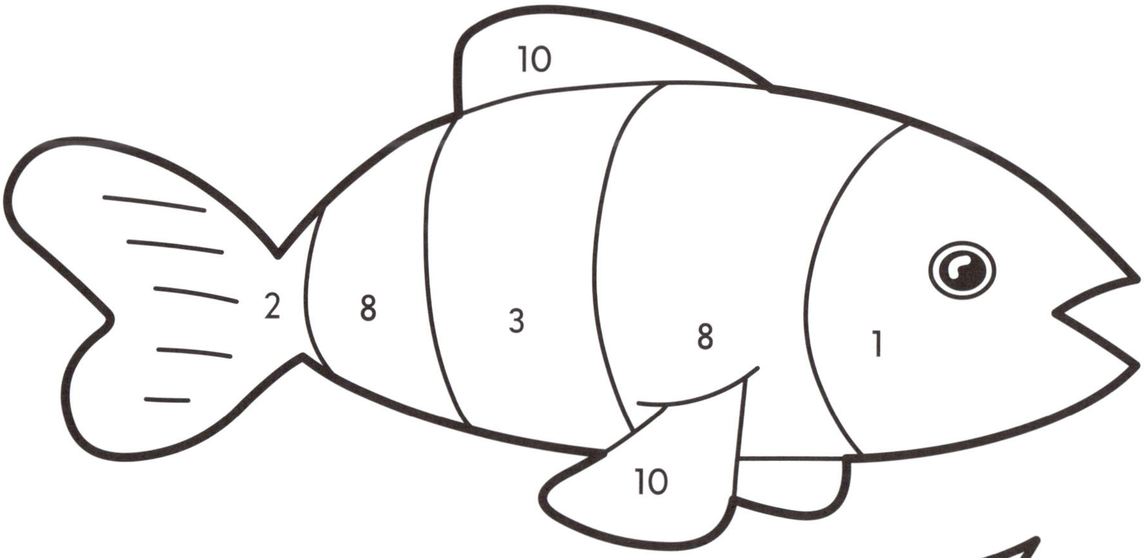

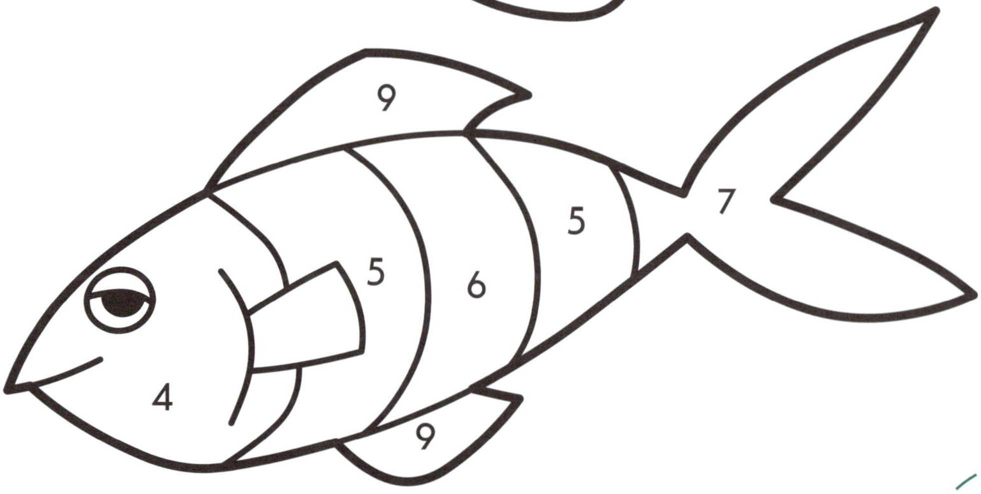

Place a sticker here.

Place a sticker here.

Well done!

You have finished the book!

**Place your last two stickers on the picture.
You can colour in the picture, too.**